IN SEARCH OF NANONOVAE

A New Theory of Element Formation

EDWARD ESKO

With Contributions by
Matthias Grabiak
David J. Nagel

IMI Press
LENOX, MA

Published in Association with

In Search of Nanonovae
Articles reprinted from *Infinite Energy* magazine

Copyright © 2020 by Edward Esko

Was Transmutation Observed at the Quantum Rabbit Laboratory?
Copyright © Matthias Grabiak
Review of Cool Fusion Copyright © David J. Nagel

ISBN-13: 978-1729780169
ISBN-10: 1729780164

Published by IMI Press
A division of the International Macrobiotic Institute (IMI)
P.O. Box 2051, Lenox, MA 01240
(413) 446-2620

Invesan.com
edwardesko@gmail.com

How likely is it that a small laboratory in New Hampshire with simplistic experimental set-ups was able to achieve nuclear reactions that are thought to be only possible under such extreme conditions as supernova explosions?

MATTHIAS GRABIAK

Contents

Introduction	xiii
In Search of Nanonovae	1
Pioneering Research	6
Was Transmutation observed in the Quantum Rabbit Laboratory?	10
Low Energy Transmutation and the Formation of Elements	26
Book Review of *Cool Fusion*	35
About the Authors	39

I hope that the book will lead to more people picking up this line of research. It is an important contribution to the development of a new paradigm of the formation of elements.

DAVID J. NAGEL

The author (right) with Woody Johnson. Photo by Alex Jack

Platinum is much rarer than both gold and silver—so rare, in fact, that all of the platinum ever mined could fit into your living room. And palladium is even rarer than that.
OUTSIDER CLUB

Introduction

In the microscopic *nanoverse*, as well as in the universe at large, no two things are identical. Common sense confirms this to be true. However, this is in contradiction to the view of modern science. In today's science, all protons are identical to one another, all electrons are identical to one another, and all neutrons are identical, as are any and all subatomic particles. It follows then that all the atoms in a given element must also be identical. All the atoms of oxygen in the atmosphere must be the same as all the others in the atmosphere; every gold atom in a ring must be identical to every other atom of gold in the ring; and any one carbon atom in a grain of brown rice will be identical to all the other atoms of carbon in that grain.

However, if we accept that nothing is identical, in reality, no two atoms or subatomic particles can ever be exactly the same. No two electrons are exactly alike. No two neutrons are carbon copies of one another. No two protons are identical. Atoms and the subatomic particles that make them up are like snowflakes, leaves, human voices, or fingerprints. They share an underlying pattern, yet each is completely unique. So the atoms and subatomic particles in Mary's gold ring are not identical to the atoms and subatomic particles in Jennifer's gold ring. Nor are any two atoms or subatomic particles in each of the rings exactly the same.

The principle of non-identity applies to all phenomena, from the microcosmic level of atoms and preatomic particles, to the macrocosmic level of galaxies, super clusters of galaxies, and so on throughout the universe.

The protons that occupy the nucleus of the atom carry a positive electric charge. The electrons orbiting the periphery are negatively charged. Neutrons carry no charge and are electrically neutral. As we know, two positively charged objects repel each other. Two negative charges will also repel. A positive charge attracts a negative charge, and vice versa. According to this law, under conditions such as those found on earth, atomic nuclei, which derive their charges from positively charged protons, will always repel each other. Nuclei are thus prevented from fusing with other nuclei.

As modern physics points out, the force of electric repulsion existing between nuclei is known as the Coulomb barrier. According to standard physics, this barrier can only be overcome under the most extreme conditions, such as extremely high temperatures, pressures, and energies. It requires energy on an order of magnitude equal to a nuclear explosion to overcome the barrier and force atoms to fuse. That is why, according to the standard model, low energy transmutation is impossible. Keep in mind however, that that assumption is predicated on the notion that all protons and other subatomic particles are identical.

However, the principle of non-identity reveals that the Coulomb barrier is relative, not absolute. If no two protons are identical, some degree of attraction, however slight, must exist between them.

It may be possible to enhance that attraction, perhaps by manipulating external conditions such as heat, pressure, and electrical energy, in a manner sufficient to weaken the barrier and allow a certain percentage of nuclei to fuse. Also throughout the universe, what has a front has a back. As the front becomes larger and more powerful, so does the back. This principle was also expressed by Isaac Newton: "Every action has an equal and opposite reaction." So, when two nuclei are repelled by one another, they are also attracted. The more strongly they are repelled, the stronger the potential for attraction.

As powerful as the repulsive force of the Coulomb barrier appears to be, existing right behind that repulsion is an equally powerful attractive force. Once again, by manipulating external conditions, it may be possible to "flip" the repulsive Coulomb force into an equally powerful attractive force, thus allowing nuclear fusion to occur. Together with the attraction of opposites and repulsion of likes, there is another process we must consider. According to the theorem: a large positive charge attracts a smaller positive charge, while a large negative charge attracts a smaller negative charge. This may help us understand how two atomic nuclei, both of which contain positively charged protons, could fuse into larger nuclei in a process of transmutation. Let us take an example from an actual experiment. As we see a nucleus of lithium, shown on the right, has 3 protons and 4 neutrons. Sulfur, at left, has 16 protons and 16 neutrons. The number of protons and neutrons determines the atomic weight of an element.

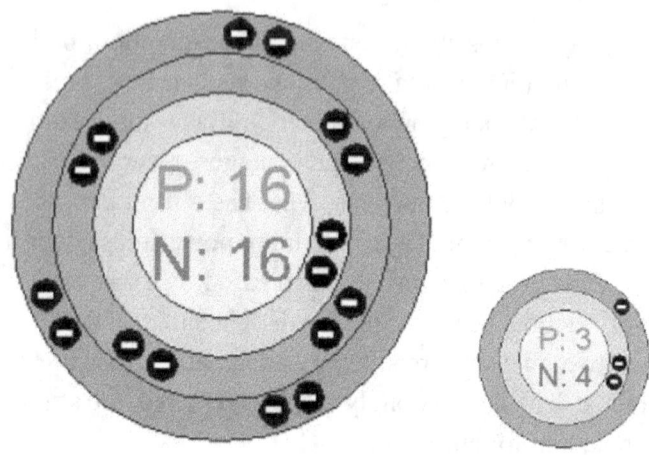

Sulfur (left) and lithium (right)

The number of electrons orbiting the nucleus determines the atomic number. The number of positively charged protons is usually equal to the number of negatively charged electrons. Thus the atom is electrically neutral. If we add the 3 protons and 4 neutrons in the lithium nucleus, we get an atomic weight of 7. If we add the 16 protons and 16 neutrons in the sulfur nucleus, the atomic weight comes out to 32. That means that a nucleus of sulfur is more than 4 times heavier than a nucleus of lithium, or that sulfur has a positive charge that is 4 times greater than that of lithium. Sulfur thus qualifies as a "large" positive, lithium as a "smaller" positive. If the theorem that big attracts small is correct, there will be attraction between lithium (small positive) and sulfur (large positive.) That attraction may offset the Coulomb barrier to some degree and explain the apparent fusion of lithium and sulfur witnessed in several experiments.

In Search of Nanonovae

In ten experiments conducted between 2009 and 2012, the Quantum Rabbit (QR) research team placed lithium and sulfur on a copper electrode under vacuum. The electrode was charged with enough electricity to generate plasma. When the experiment was finished, test material was collected and sent to an outside lab for analysis. In each of the ten experiments, analysis revealed the presence of potassium (K), although, aside from minute traces in the quartz tube, no potassium was introduced into the test.

The values for potassium ranged from a low of 3 parts per million to a high of 750 parts per million (refer to *Corking the Nuclear Genie*, by Edward Esko and Alex Jack, Amberwaves Press 2014.) The QR researchers were testing the following formula:

$^7Li + {}^{32}S \rightarrow {}^{39}K$
Lithium-7 + sulfur-32 → potassium-39

In our low energy vacuum tube experiments, we attempt to create conditions similar to those in outer space. Polarity animates the flow of energy in the tube as well as in the galaxy. Electric current passes between anode and cathode, separated by several inches. Similarly, electric current passes between the oppositely charged poles of the galaxy, separated by 100,000 light years. It may be that in this highly charged plasma sphere, elements are continually forming in a never-ending process of transmutation. It may also be that in the QR tube, also containing highly charged plasma, transmutation is taking place on an infinitely smaller scale. Microcosm equals macrocosm. As is above so is below. Skeptics often cite contamination as the source of the purported transmutation products reported in experiments. When Alex Jack, Woody Johnson and I visited Dr. Peter Haglestein at MIT several years ago, Dr. Haglestein, who has several decades of experience with LENR research, warned that critics would always cite contamination, no matter how convincing the evidence.

As Einstein famously said, "If the facts don't fit the theory, change the facts." In many of our studies, the amount of purported transmutation products is extremely small, thus results could be attributed to contamination. However, in some of our studies, the amounts of purported transmutation products, including rare metals, were simply too large to be attributed to contamination.

We may be on the cusp of a huge change in the way we view the world. A new model of element formation could be a critical step in that process. A new era may now be beginning. It may be that elements are forming through a process that is not clearly understood. The process of creation may be occurring throughout the universe at this very moment. Creation may simply be part of the peaceful and orderly process of change that is without beginning or end, and which inhabits a realm beyond time and space.

Edward Esko
Pittsfield and Lenox, MA

A continual process links the elements. They are part
of an evolutionary continuum.
MICHIO KUSHI

In Search of Nanonovae

> The abundance of palladium in the earth's crust is estimated to be about 1 to 10 parts per trillion. That makes it one of the ten rarest elements found in the earth's crust.
>
> Encyclopedia.com

In his article entitled, "Was Transmutation Observed at the Quantum Rabbit Laboratory?" (*Infinite Energy* issue 92), physicist Matthias Grabiak asked the question: "How likely is it that a small laboratory in New Hampshire with simplistic experimental set-ups was able to achieve nuclear reactions that are thought to be only possible under such extreme conditions as found in supernova explosions?" Grabiak was questioning the low energy transmutations reported to have taken place in the Quantum Rabbit laboratories from 2007 to 2013.

Grabiak is using as reference the popular theory that heavy elements such as palladium, platinum, and gold are formed in the explosions of dying stars known as supernova, or in the collision of "neutron" stars left over from supernova explosions. The latest such collision, thought to have been witnessed in 2017, produced a new star known as a "kilonova." The star faded after several days, but, according to the hypothesis, its high temperatures and pressures caused heavy elements, such as gold, to form. Astronomers believed they witnessed the formation of gold on a cosmic scale, claiming that "10-earth's" worth of gold was ejected into space.

The modern theory of element formation is based on the belief in a cosmic "chain of violence" that started with the big bang (mostly hydrogen and helium), and continued through to element creation in stars (up to iron), and peaking in supernovae, the sudden violent explosion of a star as the result of high temperature—6,000 times hotter than the sun—and pressures (iron through to uranium.)

Not only are the events in this hypothesis centered on violence, they are somewhat fixed and static, and centered on the "death" of stars, and not on the endless process of creation that is life itself. According to Britain's Astronomer Royal Sir Martin Rees, "We are literally the ashes of long dead stars."

So as to provide an alternative to the hypothesis that the elements were created in a chain of violence and death, the research team at Quantum Rabbit LLC has been conducting experiments on low energy transmutation. These experiments are attempting to duplicate on the microscopic scale events that may be occurring on the macrocosmic scale, specifically, the ongoing process of element formation. These experiments suggest it may be possible to create what we refer to as *nanonovae* (as opposed to *supernovae*):

microscopic events requiring relatively little in the way of temperature, pressure, and energy and out of which elements are formed.

In an experiment held at Continuum Energy Technologies in Fall River, MA in 2016, pure tungsten powder was placed in a graphite crucible together with small pieces of pure boron. The crucible was attached to a clip that connected it to a power pack consisting of four 12-volt batteries. A carbon rod was attached to a clip connecting it to the opposite pole of the battery pack. The rod was brought into proximity with the tungsten and boron sitting at the bottom of the open crucible. An electric spark, similar to a lightning strike, resulted. The carbon arcing process was repeated about fifty times.

Upon ICP analysis at Northern Analytical Laboratory, the residue at the bottom of the crucible was found to contain gold in the amount of 3,000 ppb (parts per billion) or 3 ppm (parts per million.) The relative abundance of gold in the earth's crust is estimated to be 0.004 ppm, substantially less than that found in the experiment. The experiment yielded a significant signal (3 ppm) to noise (0.004 ppm) ratio.

The formula being tested in the experiment was as follows:

$$^{11}B + {}^{186}W \rightarrow {}^{197}Au \text{ (boron-11 + tungsten-186} \rightarrow \text{gold-197)}$$

In another experiment, conducted at M & M Glassblowing in Nashua, NH, in 2018, the research team placed pure zinc powder in a vertical vacuum tube and mixed it with pieces of pure sulfur. The materials were placed on the surface of a copper electrode that was used as the lower anode. A copper cathode was positioned above the anode and the electrodes and materials were sealed in a quartz tube. The tube was pumped down to 3.5 torr and about 50 amps of direct current were used to electrify the anode and cathode. A glow discharge was produced, vaporizing both the zinc powder and sulfur pieces. After about 15 minutes the electricity was shut off and the tubes allowed to cool.

ICP analysis at Northern Analytical Laboratory found palladium at 2,000 ppb (2 ppm.) Palladium is an incredibly rare element. According to *Encyclopedia.com*, palladium is found in the earth's crust at only 1 to 10 parts per trillion (ppt.) If that estimate is accurate, the experiment resulted in another significant signal to noise ratio: 2,000,000 ppt versus 1 to 10 ppt. The experiment was designed to test the following formula:

$$^{34}S + ^{68}Zn \rightarrow ^{102}Pd \, (\text{sulfur-34} + \text{zinc-68} \rightarrow \text{palladium-102})$$

If elements are indeed being created in nanonovae, what process can help explain this phenomenon? The explanation may be as simple as observing what happens when we throw two stones into a pond, one after the other.

The "ripple-wave" model of low energy transmutation

The two separate ripples amplify and merge into a larger ripple. In the *nanonova* matrix, under the prompting of directed heat and energy, nuclei may undergo a similar transformation. Two lighter nuclei amplify and expand, dissolving their identities and merging into a new larger and heavier nucleus.

Pioneering Research

We trace our lineage in the field of low energy transmutation (see *Cool Fusion* and *Corking the Nuclear Genie*, Amberwaves Press 2012 and 2014) to the work conducted by George Ohsawa, Louis Kervran, and Michio Kushi in the 1950s and 1960s, in Tokyo, France, and Cambridge, Mass. However, readers might be interested to know that modern research on low energy transmutation actually took place four decades earlier at Tokyo University in Japan. Coincidentally, Tokyo University is the Alma matter of Michio Kushi, who, in the 1970s, introduced the Quantum Rabbit partners to low energy transmutation through his Boston lectures. In 1924 and 1925, Professor Hantaro Nagaoka and his associates at Tokyo University conducted some 200 experiments.

Nagaoka used high-current electric arc discharges between tungsten electrodes immersed in liquid hydrocarbon transformer oil in which they detected the successful transmutation of tungsten into visible flecks of gold and platinum. In June 1925, Nagaoka went a world tour in which he spoke to scientific and lay audiences about his transmutation experiments and handed-out samples with tiny bits of gold that had been created.

A letter on his work was published in the journal *Nature*

in July 1925, in which he encouraged other scientists to try to duplicate his results.

Unfortunately, modern nuclear physics preempted Nagaoka's work and his results were not followed up on, until the independent work of Ohsawa, Kervran, and Kushi four decades later. The Quantum Rabbit research of the early 2000s was based on this independent work.

If valid it seems Prof. Nagaoka achieved transmutation of tungsten into gold through a process of low energy *fusion*. Before hearing of Nagaoka's work, I developed a formula for this possible low energy fusion reaction: $^{11}B + {}^{186}W \rightarrow {}^{197}Au$ (boron-11 + tungsten-186 $\rightarrow$ gold-197.) In experiments conducted in 2008, 2009, and 2011, the QR team apparently produced nano-quantities of gold through a process of low energy *fission*.

Our experiments involved the low energy fission of lead into gold through the following formula: $^{204}Pb \rightarrow {}^{7}Li + {}^{197}Au$ (lead-204 $\rightarrow$ lithium-7 + gold-197.) In our theory, the low energy fission of lead into lithium and gold is triggered by a low energy fusion reaction: $^{7}Li + {}^{32}S \rightarrow {}^{39}K$ (lithium-7 + sulfur-32 $\rightarrow$ potassium-39.) In all experiments, potassium was detected in the end product. However, unlike the results reported by Nagaoka, the microscopic quantities of gold in the QR tests were not visible to the naked eye, but were detectable only through inductively coupled plasma spectroscopy (ICP.)

Invesan Technologies is current working on proposals for a new round of experiments on low energy transmutation, including studies of foundation formulas, rare earth metals, rare metals, and remediation of nuclear materials. We welcome inquiries and are seeking venture partners. We would like to acknowledge Prof. Nagaoka's pioneering work and hope to confirm and develop it.

> The experiments described here are done in simple laboratories with a low budget, yet are significant because they show with little doubt that elements can be transmuted at temperatures far below what science would consider possible.
>
> BILL ZEBUHR

Matthias Grabiak

WAS TRANSMUTATION OBSERVED IN THE QUANTUM RABBIT LABORATORY?

It has been over twenty years now since Pons and Fleischmann claimed to have observed evidence of nuclear reactions occurring at low energies in a tabletop experiment, reporting an output of excess heat that could not easily be explained as the result of any chemical processes.[1] If such a finding could be confirmed it would hold enormous promise for future technologies, possibly providing a clean and safe solution for the world's energy needs. But soon after the announcement by Pons and Fleischmann, other researchers failed to reproduce their results, and the majority of scientists came to consider "cold fusion" dead.

Nevertheless, there is a small minority of scientists still performing experiments in the field of LENR (low-energy nuclear reactions), a newer term that was introduced later to replace "cold fusion." LENR experimentalists have made various claims of success, but these have not been sufficient to provide evidence so compelling that it would be generally accepted.

There are reports of excess heat in newer experiments using better measuring techniques, but this cannot be reproduced with sufficient consistency and the observed heat is only on the order of magnitude of the heat entering the experiment—not enough to heat a cup of tea and not enough to convince skeptics. Evidence of particle tracks in CR-39 plastic sheet detectors that may have been produced as a result of nuclear reactions has been obtained at a U.S. Navy laboratory,[2] but even this evidence has not been strong enough to convince the majority of scientists.

The most direct proof of nuclear reactions would be the appearance of substances that were not originally present, strongly suggesting transmutation of nuclear elements, provided that contamination can be ruled out. Most experiments are focused on the reaction of deuterium or ordinary hydrogen nuclei, which would produce helium or tritium. But at least as far as He^4 is concerned it occurs naturally in our atmosphere, making it hard to rule out contamination.

A more promising indicator would be the production of heavier elements involving larger nuclei, as long as large enough concentrations could be obtained. In addition, these transmutation products would surely have isotopic compositions[3] that deviate significantly from the ones found in natural occurrences of the elements. This could be just the evidence that supporters of LENR would need to convince a large majority of scientists.

There are some experimentalists who claim to have observed such unusual isotopic compositions. But as Michael Schaffer[4] wrote in a 1999 article for *Scientific American* about the status of low-energy nuclear reactions, "Production of such heavy nuclei is so unexpected from our present understanding of low-energy nuclear reactions, that extraordinary experimental proof will be needed to convince the scientific community." As of today, that has not been achieved.

Enter Quantum Rabbit. Quantum Rabbit is a small laboratory in Nashua, New Hampshire run by Edward Esko, Alex Jack and Woodward Johnson. Remarkably, they started their experiments without any special expertise in nuclear physics. They have performed various experiments where they have seen the anomalous appearance of different substances, which they suggest may be a hint of transmutations, that is, changes in the nuclear structure of their test substances.

In their most promising experiments, the purported transmutation products showed up in concentration in the order of magnitude of thousands parts per million within the materials used in the experiments, as measured reliably by outside laboratories. In order to rule out the possibility that the detected substances could have been present before the experiment, certified pure samples were used for all the test materials and the test tubes used were carefully examined for contaminants as well.

The best effort was made to avoid contamination throughout the experiment. Various kinds of test materials, like lithium or sulfur, were placed in a vacuum tube between two metal electrodes. The air was then pumped out of the tube and oxygen was pumped back reaching a pressure of a few Torr. Electricity was then applied to the electrodes and the test material was heated until it started evaporating. Both electrodes, as well as the test materials, were then sent to a laboratory where they were tested for traces of various elements that should not have been present at the beginning of the experiment.

Interestingly enough, the elements that were detected afterward could well be explained as fusion products of the elements that entered the experiment. For example, using iron electrodes with lithium as a test material, copper was found after the experiment.

A quick look at the periodic table confirms that a copper nucleus would result from simply combining a nucleus of iron and lithium, without additional reactions like beta decay or electron capture that would further change the end products. I have compiled an excerpt of some of their tests results in Table 1. The suggested transmutation product was commonly found both on the test substance residues and on the electrodes; the value listed is the higher one observed.

In another type of experiment, non-metallic graphite or silicon powders (scientific grade 99.999% pure) were placed in a pure (99.999%) graphite crucible. The powders were charged with 36 volts of direct current through a pure (99.999%) graphite rod. The crucible was connected to the negative pole, the rod to the positive pole of a power pack consisting of three 12-Volt solar-charged batteries.

Table 1. Some Quantum Rabbit Test Results

Fe + Li → Cu   1500 ppm (anode)
Stainless electrodes with lithium test substance

Zn + S → Pd   91 ppm (sulfur residue)
Copper/zinc electrode with sulfur test substance

Zn + O → Sr   14 ppm (anode)
Copper/zinc electrode with sulfur test substance

S + O → Cr   198 ppm (sulfur residue)
Copper/zinc electrode with sulfur test substance

Ag + Li → Sn   3 ppm (lithium residue)
Copper/silver electrode with lithium test substance

Cu + Li → Ge   2190 ppm (cathode)
Copper electrodes with lithium test substance

C + O → Si   138 ppm (cathode)
Graphite electrodes with sulfur test substance

Fe	iron	O	oxygen
Li	lithium	Sr	strontium
Cu	copper	Cr	chromium
Zn	zinc	Ag	silver
S	sulfur	Sn	tin
Pd	palladium	Ge	germanium

The powders received between 100 to 200 strikes from the charged rod. Upon cooling the powders were tested for magnetic properties with a neodymium magnet before packaging and shipping them to an outside lab for EDS and ICP analysis.

Very clearly the treated graphite powders showed magnetic properties, which could point to the possibility that magnetic metals like iron, cobalt or nickel were produced by transmutation of the carbon. However, one needs to be aware of the fact that carbon itself was found in 1997 to have an allotrope (a specific form of an element, *e.g.* graphite and diamond are the more common allotropes of carbon), carbon nanofoam, that turned out to exhibit ferromagnetic behavior like iron.[5] Thus evidence of magnetism alone is no proof that it is caused by iron, cobalt or nickel resulting from transmutation. Nonetheless, the chemical analysis of the graphite powder shows 4700 ppm iron, along with other elements, notably silicon at 1.5%.

One might think that the concentrations of putative transmutation products seen in the Quantum Rabbit experiments could be large enough to be considered a giant breakthrough in the field of LENR, but so far these experimental results have garnered very little attention.

These results are so contrary to established theory that it is extremely hard to accept them, much harder than most other LENR evidence that is focused on reactions between hydrogen isotopes like deuterium. For deuterium—having only one positive charge each—the mutual electric repulsion between nuclei known as the Coulomb barrier is as small as it can get. Still, even for deuterium it seems extremely difficult to explain theoretically how this repulsion can be overcome to accomplish nuclear reactions in systems with low energy and low density. For most scientists it therefore sounds like suggesting that an ant is capable of pushing a baby stroller, making it hard for LENR to find general acceptance. The idea of nuclear reactions between heavier nuclei as supposedly observed in the Quantum Rabbit experiments goes far beyond that—it is more like suggesting that an ant is pulling a freight train. The electric repulsion becomes so forbiddingly strong that such reactions at low energies would be thoroughly ruled out by the known laws of physics, particularly by the laws of electricity that have been extremely well established and extremely well confirmed.

Conventional physics tell us that it requires pressures and temperatures comparable to those of the interior of the sun to overcome the Coulomb barrier even for the very lightest elements in order to achieve fusion. But, for fusing heavier elements as suggested by the Quantum Rabbit experiments, only cataclysmic scenarios like supernova explosions will suffice.

When only singly charged nuclei like ordinary hydrogen or deuterium are involved, one may try to think of some extraordinary mechanism that could explain how the charge of the nucleus may be neutralized by an electron, like the temporary formation of a deflated state,[6] or even by wildly speculating that a proton and an electron can temporarily combine to form a neutron[7]—just to name two examples without giving any credence to any particular theory (see my criticism of Widom and Larsen[8]). The point is, finding such a mechanism for singly charged nuclei like ordinary hydrogen or deuterium is already extremely difficult, and none of the existing explanations are entirely convincing to say the least. But when it comes to nuclei with higher charges it would require some collective effect involving many electrons at once to neutralize the entire charge of the nucleus, which seems all but impossible.

The most obvious conclusion is that transmutation can be ruled out as an explanation for the results reported by Quantum Rabbit, as nuclear reactions at such low energies strongly contradict the established theories of physics. One might ask though if this could be a hint that the established theories are wrong, or at least if there may be some entirely new physical phenomena involved that cannot be explained within the current theories. After all, it often turns out that discrepancies between existing theories and experimental results prove to be gateways to revolutionary changes in our understanding, bringing about radical paradigm shifts.

Would the experiments at Quantum Rabbit be able to bring about such a radical change? Most scientists would consider that highly unlikely, probably even most researchers in the field of LENR.

Furthermore, LENR as proposed by Quantum Rabbit is most at odds with the extremely well established laws of electromagnetism worked out by Faraday and Maxwell in the nineteenth century, as it is the immensely strong electric repulsion that would prevent the nuclei from coming close enough together to allow for a nuclear reaction. And even if one could devise a theory that can explain how that repulsion can be overcome, it still has to be consistent with all other evidence.

The new theory must not predict nuclear reactions to be happening much more easily and frequently in other scenarios than is actually observed, especially if it would entail catastrophic consequences. For example, it may be difficult to explain why the hydrogen in the sun does not immediately get converted into heavier elements by a giant explosion instead of getting it slowly consumed over billions of years. But beyond that it may be hard to explain why transmutations are not easily observed as a commonplace occurrence in many other experiments.

While the experimental results may look quite intriguing, the researchers at Quantum Rabbit are well aware that there are still possible sources of contamination that have not been properly addressed. They agree that going forward in this line of research their experiments would have to be repeated at a research institute or an established laboratory.

One could take the viewpoint that the arguments against LENR of heavier elements are so compelling that transmutation has to be ruled out, and therefore any attempt to look deeper into the matter would be a waste of time and resources. This would be similar to the highly skeptical viewpoint that most scientists have towards LENR research in general. On the other hand, what is special about the experiments at Quantum Rabbit is the fact that they have been achieved with relatively modest means.

It might therefore be well worth the effort to try to reproduce at least some of their results, focusing on obtaining stronger evidence in favor or against transmutation with some easily performed experiments. There are some obvious steps that could be taken next. First and foremost, it would be quite revealing to test the isotopic composition of the putative transmutation products. Isotopes are variations of nuclei that have the same number of protons and thus behave the same chemically, as it is their charge that determines their interaction with the electrons, but they differ in the number of neutrons in their nuclei. Most elements consist of a mix of various isotopes, and when found in nature there is very little deviation in their relative abundance. This is often used as some kind of fingerprinting of materials, as their compositions differ ever so slightly when they come from different origins. When produced by transmutation, however, one would expect radical deviations from the ordinary mix.

For example, if we look at the suggested reaction Fe (iron) + Li (lithium) → Cu (copper) in detail we find that copper found in the environment is made up of about 69% Cu^{63} consisting of 29 protons + 34 neutrons = 63 total, and 31% Cu^{65} with 29 protons + 36 neutrons = 65 total. Lithium consists mostly Li^7 (3 protons + 4 neutrons) with an abundance of about 92.5% and of Li^6 (3 each) at 7.5%.

The only simple fusion reaction yielding Cu^{65} would therefore be $Fe^{58} + Li^7 \rightarrow Cu^{65}$, since Fe^{59} is not stable and does not occur in nature, ruling out the reaction $Fe^{59} + Li^6 \rightarrow Cu^{65}$. But natural iron contains only 0.28% Fe^{58}, consisting mostly of Fe^{56} with 92% and some other isotopes lighter than Fe^{58}. One would therefore expect very little Cu^{65} from the transmutation reaction. If the analysis of the copper isotopes found in the experiment showed a fairly large proportion of Cu^{65} anywhere close to the natural 31% it would be a clear indication that the copper detected was not a result of transmutation, but most likely came out of the environment as contamination. If on the other hand the analysis of isotopes showed an apparent lack of Cu^{65} it would bolster the claim of transmutation considerably.

Similar tests of the end products could be performed for many of the other experiments. If the silicon found in the graphite experiment was obtained by transmutation of lighter elements, like carbon + oxygen or even nitrogen + nitrogen from the atmosphere, one would expect it to be almost exclusively Si^{28}, with considerably less than 4% Si^{29} that is found in naturally occurring silicon.

Analyzing the nickel found in the same experiment could provide even more valuable clues. Nickel naturally consists of a pronounced mix of isotopes, mostly Ni^{58} with 68%, Ni^{60} with 26%, and Ni^{62} with 3.6%, which may make it easier to observe clear deviations in the isotopic composition.

Future experiments could be devised with the goal of maximizing the evidence obtained by analyzing the isotopic composition of the supposed transmutation products. One of the most promising examples could be to repeat the Quantum Rabbit experiments using manganese + lithium in an attempt to produce nickel. Manganese has only one stable isotope, Mn^{55}, and lithium is 92.5% Li^7 and 7.5% Li^6, thus one would expect mostly Ni^{62} and some Ni^{61} from a simple fusion of manganese and lithium nuclei. But these two isotopes make up only less than 4% of naturally occurring nickel. Ordinary nickel on the other hand consists of 94% of the lighter isotopes Ni^{58} and Ni^{60}. This should make it obvious whether any observed nickel was produced by transmutation or originated from contamination. Even if more complicated reactions could occur turning Ni^{62} and Ni^{61} into lighter isotopes, for example by neutron emission, the isotopic composition of any nickel obtained by transmutation would likely deviate considerably from that of ordinary nickel, thus providing conclusive evidence.

A very important point to note about the isotopic composition is the effect that contamination has on the evidence. Simply testing for concentrations of various elements in treated test samples as has been done so far has the problem that contamination can result in false evidence.

With respect to isotopic composition, on the other hand, contamination would diminish the evidence, as any contaminants from the environment would have the naturally occurring mix of isotopes. Thus, if the analysis of the end products were to show a highly anomalous isotopic composition, it would be strong evidence for something unusual going on, warranting further research.

Some other simple and straightforward tests could be performed as well. It would be simple enough to test for particle radiation by inserting a cheap CR-39 plastic polymer, a technique often used in other LENR experiments.

If any hints of radiation are found it could be taken as a possible indication that nuclear reactions have indeed occurred. With a little bit of creative thinking one can come up with other tests that could be simple enough to perform. So far the researchers at Quantum Rabbit have been more focused on testing for elements that would be expected from nuclear reactions of the substances involved in the experiment—but what about looking for elements that would not be expected as end products of nuclear reactions? For example, while iron + lithium would be expected to yield copper, what about using either cobalt + lithium or nickel + lithium in the experiment instead, with cobalt and nickel being the elements following iron in the periodic table?

Would one still see copper as an end product, maybe in similar concentrations as in the iron experiment? If so it may be an indication that the copper detected after the experiment resulted from some contamination and not from transmutation, which should result in zinc or gallium instead that come after copper in the periodic table.

How likely is it that a small laboratory in New Hampshire with simplistic experimental set-ups was able to achieve nuclear reactions that are thought to be only possible under such extreme conditions as found in supernova explosions? If low-energy nuclear reactions between heavier, more highly charged nuclei are possible, why was that not detected and firmly established long ago by generally accepted experiments? The arguments against transmutation seem compelling and it may seem inevitable to conclude that it has to be ruled out, contamination being a much more likely explanation. The researchers at Quantum Rabbit themselves admit to that possibility. However, as remote as the possibility of LENR of heavier elements may seem, if it was confirmed it would have earth-shaking consequences, both theoretically and practically. Being in conflict with the current theories of physics, it would possibly pave the way for a new revolution that would likely bring about a completely new understanding of physics.

It would surely open the door for a huge number of applications that could introduce radically new industries, like producing iron from coal or copper from iron and lithium, to name just two examples.

Further exploration could well uncover new inexhaustible sources of energy from nuclear reactions that could be exploited in safe and peaceful ways. Even if this all seems absolutely farfetched, it may well be worth the extra efforts to perform a few more simple experiments that can either confirm low energy nuclear reactions of heavier nuclei in these types of experiments, or rule them out. The result may well turn out to be negative as conventional physics would predict, but if so, we will have made sure that we are not passing up a giant opportunity, and we will be able to agree that the established laws of physics have withstood another challenge.

References
1. Fleischmann, M., Pons, S. and Hawkins, M. 1989. "Electrochemically Induced Nuclear Fusion of Deuterium," *Journal of Electroanalytical Chemistry*, 261, 2, Part 1, April 10.

2. Anderson, M. March 25, 2009, "New Cold Fusion Evidence Reignites Hot Debate," http://spectrum.ieee.org/energy/nuclear/new-cold-fusion-evidence-reignites-hot-debate
3. A kind of fingerprinting of materials that will be explained in more detail later in this article.
4. October 21, 1999, "What is the Current Scientific Thinking on Cold Fusion?" http://www.scientificamerican.com/article.
5. Rode, A.V. *et al.* 2004. "Unconventional Magnetism in All- Carbon Nanofoam," *Phys. Rev. B*, 70: 054407.
6. Heffner, H. 2008. "Deflation Fusion: Speculations Regarding the Nature of Cold Fusion," *Infinite Energy*, 14, 80, 38-46.
7. Widom, A. and Larsen, L. 2005. "Ultra Low Momentum Neutron Catalyzed Nuclear Reactions on Metallic Hydride Surfaces," *Eur. Phys. J. C*, March 9.
8. newenergytimes.com/v2/sr/WL/GrabiakCritique-Widom-LarsenFeb4-2010.pdf

It is premature to reduce the vital process to the quite insufficiently developed conceptions of 19th and even 20th century physics and chemistry.
L. DE BROGLIE

Low Energy Transmutation and the Formation of Elements

The following is from e-mail correspondence between Matthias Grabiak (Was Transmutation Observed at the Quantum Rabbit Laboratory?) and Quantum Rabbit founder Edward Esko.

Esko: What if the new theory of element formation wasn't a theory at all, but an entirely new paradigm, a paradigm shift, so to speak? What I'm proposing may be an entirely new context that encompasses the existing theories but from a new angle.

The new angle applies as much to elemental, atomic, nuclear, and sub-nuclear forces as it does to gravity and the movement of the stars and planets. Within this new cosmology, the varied theories of science are embraced and explained anew. The key is the origin of the force holding the nucleus together and the force that makes objects fall to the surface of the planet. My view is that this force (which takes many forms, including magnetism, electricity, gravity, etc.) doesn't originate from the body in question, be it a nucleus or a planet, but from infinite space. It is not a pulling force but an infinitely pushing force.

Grabiak: Paradigm shifts are very common in physics, and they usually break new ground without pushing aside the existing theories completely. Newton's laws of mechanics are still extremely useful, even after Einstein's theories of relativity and quantum mechanics have proven it wrong. This is usually a question of what scenarios you are dealing with. Relativity comes into play when dealing with objects moving close to the speed of light, and quantum mechanics when dealing with objects at atomic distances, for anything within our daily experience Newton's law of mechanics work quite excellently. The problem I see, if you are encompassing the existing theories, it would imply that the conclusions reached from those theories are still valid. Among other things this means that the Coulomb barrier is still there, thus the new paradigm does not help to provide a new explanation for LENR.

Esko: Let me try to sum up:

1. The attractive force between opposite charges is generated from the outside, so that they are being pushed together like opposite magnets.
2. Elements placed between the two charges will undergo changes, depending upon the degree of

charge and other conditions, for example, pressure; vacuum versus atmosphere.
3. Lighter elements may fuse to form heavier elements; heavier elements may fission to form lighter elements, at least on the nano-scale.
4. The tendency of all matter is to condense into an infinitesimal point due to the force generated by the surrounding quantum vacuum. The reason this doesn't occur is because of the opposite, electrical repulsive force that exists between like charges, for example, proton and proton.
5. By manipulating pressure toward the quantum vacuum, and altering, even slightly, the plus or minus charge of two nuclei relative to each other, it may be possible to overcome the electrical repulsive force to the extent of allowing a small percentage of nuclei to bunch together to form a heavier nucleus. The quantum vacuum holds the nuclear particles of the new heavier nucleus together.
6. This process may also be occurring on the macrocosmic scale at the level of galaxies. Galaxies are the matrices in which matter is formed in the universe.
7. Observing the galaxy from the side, the force converging toward the center from the left represents one of the charges (anode) and the force converging from the right, the other charge (cathode.)

Hydrogen and helium are continuously being formed from the condensation of energy into electrons and protons. These elements make up the primordial gaseous cloud that gives rise to the galaxy. These elements are squeezed toward the center by the quantum vacuum and, once they arrive there, are energized by the two oppositely charged vectors, appearing as two converging lines of force.

It is here that these light elements are forced outward toward the periphery, while undergoing nuclear transmutation into the heavier elements. As nuclear and elemental material is ejected from the center of the galaxy toward the outer spiral orbits, billions of tiny condensations occur in which the process of nuclear transmutation continues through the periodic table, arriving at the radioactive elements. These tiny condensations represent stars and solar systems.

This process is taking place throughout the universe and has no fixed origin from the macrocosmic perspective. Each galaxy, however, has a defined beginning and end point that can be measured from our relative perspective. If you look at a photo of a QR vacuum tube with element plasma inside, it bears a striking resemblance to the galaxy side view. Perhaps we are, on the microscopic level, replicating a process that takes place continually on the galactic plane.

Upper: Side view of the Milky Way Galaxy. Lower:
Helium plasma in QR vacuum tube

This new paradigm replaces the big bang and other creationist interpretations of the genesis of the universe, including non-scientific interpretations.

Grabiak: I would still like to stress that with point 5, you are in complete contradiction with the well-established laws of electromagnetism, postulating a radical change like that means that your theory will likely make lots of other predictions that contradict existing evidence, likely untangling your theory before it even comes together. It is absolutely not as you suggested that your theory encompasses the existing theories, it is going full out on a collision course.

Question, is it right that the universe knows to push opposite charges together and like charges apart? Can you provide any details of how that works? I have a few more thoughts. How about Einstein's two equivalent viewpoints of gravity? When observing a falling elevator from outside we would say the objects inside are accelerated towards earth by gravitational forces, but an observer within the elevator that is accelerated at the same rate as the objects inside would see the elevator as a chamber with no gravitational field inside and no forces acting on the objects.

Can you accommodate that within your viewpoint? Is the universe pushing or is it not pushing? That example can also be turned around. Let's say there is an elevator that has been abducted into outer space, and an alien being starts pulling it and accelerating it. From the inside of the elevator it looks like a gravitational force with objects appearing to fall in the direction away from the pulling alien, but an outside observer would say that there is no force at all acting on the objects inside. Again, is the universe pushing the objects or is it not pushing?

Question, why has the concept of local field theories been so successful guiding us towards theories that have been very successful explaining nature? All current theories are of that nature, gravity, electromagnetism, and strong and weak interactions.

The idea is that the behavior of particles and fields is only determined by their immediate surroundings. An electric particle reacts to the electric field at its position, no matter if that field is caused by a charged particle on one side, or an oppositely charged particle on the opposite side, whether the field is caused by a small charge nearby or a big charge far away.

This appears to be in stark contrast to your ideas. So, if your ideas are correct, can you think of a way of showing that local field theories are wrong, and a faraway charge in the universe does have an effect? (Side question, how would the effect of the universe be communicated?)

The concept of local field theory works well with the idea of "signals" (e.g. field waves) that travel at finite speed). On the other hand, if you were encompassing the existing theories, you would have to explain why the paradigm of local field theories is not a contradiction to your paradigm.

Esko: In the final analysis, it's premature for us to be speculating as to a theory to explain the QR low energy transmutation results. Most observers wouldn't accept the QR data to begin with. Our methods and controls were admittedly impromptu, makeshift, and not well controlled. The key objection—that of contamination beforehand has not been adequately addressed.

We have, for example, been using the Certificates of Analysis provided by the supplier for the control values of the various elements used in the tests, rather than actual analysis of the control samples beforehand. Contamination could have crept into our results from myriad angles, from the handling of the materials prior to and after the tests, to the composition of the electrodes and glass tube materials, to the air quality in the room.

Given our level of sophistication, we can't say for certain whether or not low energy transmutation is in fact taking place, or is in fact real. However, my associates and I have seen results that seem to indicate that something unexplained is taking place in these tests. That something has occurred with such regularity and predictability that it strongly suggests our theory of low energy transmutation is valid and merits further testing under stricter controls. My point is that our evidence is not solid enough to challenge the current scientific paradigm. I would feel more comfortable if a university such as MIT or Berkeley, or an established lab such as Cavendish or Oak Ridge, ran dozens of tests that validated our results. In other words, until low energy transmutation is established as fact, without a doubt, it is premature to debate the how or why it is or is not taking place.

If it turns out that low energy transmutation is confirmed beyond reasonable doubt, the great theorists of that time will do what they have always done, improvise a theory to fit the apparent facts. This may or may not occur during our lifetimes. I'd like to propose that we hold off on trying to come up with a theory for our results until more evidence comes in. The evidence will either add more proof to the existing paradigm, or force some type of revision or modification in it. I suggest we instead try to garner support, both scientific and financial, for another round of tests, using the published QR reports as the starting point.

We are probably on the eve of a great change in physics.
C. LOUIS KERVRAN

David J. Nagel

BOOK REVIEW OF
COOL FUSION

It is good to have the book *Cool Fusion*. Earlier, I read articles about the authors' experiments in *Infinite Energy*. Now, it is really nice to have the entire body of work in one well-organized document.

By way of background, especially for those who missed the articles in this magazine, Quantum Rabbit LLC (QR) was founded in 2005 by Edward Esko, Woody Johnson, and Alex Jack. They are educators and independent researchers living in Massachusetts. Their purpose is to continue the experimental LENR work of Louis Kervran, George Ohsawa and Michio Kushi.

Between 2005 and 2009, QR conducted over twenty-five experiments on LENR at small labs in Nashua NH, Owls Head ME and Bellows Falls VT. At the laboratory in Bellows Falls, QR conducted open-air carbon-arc studies based on the experiments by Ohsawa and Kushi in the mid-1960s.

Esko and Jack confirmed the results of the two earlier researchers. They reported the creation of magnetic graphitic powder, plus the anomalous appearance of other elements after the experiment: silicon at up to 10,500 ppm, magnesium at 1800 ppm, iron at 4700 ppm, aluminum at 7800 ppm, titanium at 440 ppm, cobalt at 160 ppm, and nickel at 1120 ppm.

Vacuum discharge experiments were conducted at the laboratories in Nashua and Owls Head. Using highly simplified methods, QR reported the anomalous appearance of a variety of metals. (The vacuum discharge results are summarized in Matthias Grabiak's article.) *Cool Fusion* describes each of these experiments in detail, with photos and illustrations showing design of vacuum tubes, electrode configurations, and test substances. This book provides a description of how the QR project came into being. The book also presents the phenomenon of cool fusion in context of ongoing research into both hot and cold fusion.

For years I have been grumping that much of the literature on transmutations does not include proper chemical analyses before and after experiments. The QR team is an exception. But, I would like to see the analytical data from all experiments presented in three-column tables.

The first column would be the element, the second its starting concentration, and the third the final concentration, with as many rows as needed to cover all the elements. Of course, the same analytical method would have to be used both before and after experiments.

Cool Fusion offers some comments on repeatability of experiments. However, it should be more quantitative, maybe giving the element-by-element enhancements for a series of experiments, one after the other, again in tabular form. There has been so much written on the reproducibility of heat measurements.

The same concern applies to transmutation measurements. The proper analyses of many elements at low levels are expensive, and there is never enough money and time to do an exhaustive job. But I favor doing fewer experiments more thoroughly rather than doing many less thoroughly.

The scientists at SPAWAR, who do co-deposition experiments, and other LENR researchers doing a wide variety of experiments, have been criticized for not repeating their work adequately. I feel that the same comment can be leveled at much of what is in *Cool Fusion*. It would be ideal if they chose one of their many experiments and ran it, say, ten times, and reported the results for each run. The use of graphite to produce Si and other metals might be a good candidate, since very pure graphite is available.

I would like to see detailed time traces of voltage, current, power, temperature, and light intensity (at various wavelengths) from the beginning to the end of these experiments. I know that obtaining them takes a lot of equipment (money) and effort. But they are really fundamental diagnostics. Only in the table on page 101 was there any indication of the applied voltage and current. I spent a couple of decades as a spectroscopist, so I would like to see spectral data, which would provide both intensities and temperatures.

The use of a template for each of the chapters in *Cool Fusion* reporting on experiments is very convenient and informative. It is easy to see what was done and found, even though I called for more details above. I also enjoyed the interesting historical summaries. One was the pioneering work of Norman Lockyer, the discoverer of helium, founder of *Nature*, and experimentalist who reportedly converted copper into calcium in vacuum studies in 1878. Another treated Sir William Ramsay, who discovered neon, krypton, and xenon, and won the Noble Prize for Chemistry. He reported in 1907 that he had transmuted copper into lithium. I knew much of what was written, but it is nice to have summaries of the bases for this work.

I hope that the book will lead to more people picking up this line of research. It is an important contribution to the development of a new paradigm of the formation of elements.

ABOUT THE AUTHORS

Edward Esko, the founder of Quantum Rabbit LLC and Invesan Technologies, designed the experiments outlined in *Cool Fusion* and *Corking the Nuclear Genie*. He is the founder of the Quantum Research Institute, a global initiative to further the study and application of low energy transmutation.

Matthias Grabiak grew up in Germany and got his Ph.D. in theoretical physics in 1988 at the Johann Wolfgang Goethe Universität in Frankfurt. From 1989 through 1992 he was a post-doctoral researcher at the Nuclear Physics Group of the Lawrence Berkeley Laboratory. He then began working in the software industry and is currently employed with Oracle, while still having a continued interest in physics.

David J. Nagel is a Research Professor at The George Washington University and CEO of NUCAT Energy LLC, a consulting company on LENR. He has been active in LENR research since the 1989 Fleischmann-Pons announcement.

BOOKS ON LOW ENERGY TRANSMUTATION

Available at Amazon.com

www.ingramcontent.com/pod-product-compliance
Lightning Source LLC
Chambersburg PA
CBHW071433220526
45469CB00004B/1511